Dangerous Snakes

Rattlesnakes

Serpientes peligrosas

Serpientes de cascabel

Written by / Escrito por: Tracy Nelson Maurer
Translated by / Traducción de: Sophia Barba-Heredia

TABLE OF CONTENTS

ÍNDICE

A Crabtree Seedlings Book
Un libro de El Semillero de Crabtree

Crabtree Publishing
crabtreebooks.com

School-to-Home Support for Caregivers and Teachers

This book helps children grow by letting them practice reading. Here are a few guiding questions to help the reader with building his or her comprehension skills. Possible answers appear here in red.

Before Reading:

- What do I think this book is about?
  - *I think this book is about rattlesnakes.*
  - *I think this book is about why rattlesnakes make rattling sounds.*
- What do I want to learn about this topic?
  - *I want to learn how to stay away from rattlesnakes.*
  - *I want to learn where rattlesnakes live.*

During Reading:

- I wonder why...
  - *I wonder why rattlesnakes don't like cold weather.*
  - *I wonder why some people eat rattlesnakes.*
- What have I learned so far?
  - *I have learned that rattlesnakes must shed their skin to grow.*
  - *I have learned that they add a ring to their rattles each time they shed their skin.*

After Reading:

- What details did I learn about this topic?
  - *I learned that rattlesnakes eat mice, birds, and other snakes.*
  - *I learned that rattlesnakes don't like cold air.*
- Read the book again and look for the vocabulary words.
  - *I see the word **rattler** on page 5 and the word **scales** on page 4. The other glossary words are on pages 22 and 23.*

Apoyos de la escuela a los hogares para cuidadores y maestros

Este libro ayuda a los niños en su desarrollo al permitirles practicar la lectura. Abajo están algunas preguntas guía para ayudar al lector a fortalecer sus habilidades de comprensión. En rojo hay algunas opciones de respuesta.

Antes de leer:

- ¿De qué pienso que tratará este libro?
  - *Pienso que este libro es sobre serpientes de cascabel.*
  - *Pienso que este libro es sobre por qué las serpientes de cascabel hacen sonidos de cascabeleo.*
- ¿Qué quiero aprender sobre este tema?
  - *Quiero aprender cómo mantenerme lejos de las serpientes de cascabel.*
  - *Quiero aprender dónde viven las serpientes de cascabel.*

Durante la lectura:

- Me pregunto por qué...
  - *Me pregunto por qué a las serpientes de cascabel no les gusta que haga frío.*
  - *Me pregunto por qué algunas personas comen serpientes de cascabel.*
- ¿Qué he aprendido hasta ahora?
  - *Aprendí que las serpientes de cascabel deben mudar su piel para crecer.*
  - *Aprendí que ellas agregan un anillo a su cascabel cada vez que mudan la piel.*

Después de leer:

- ¿Qué detalles aprendí de este tema?
  - *Aprendí que las serpientes de cascabel comen ratones, pájaros y a otras serpientes.*
  - *Aprendí que a las serpientes de cascabel no les gusta el aire frío.*
- Lee el libro de nuevo y busca las palabras del vocabulario.
  - *Veo la palabra **escamas** en la página 4 y la palabra **cascabel** en la página 5. Las demás palabras del vocabulario están en las páginas 22 y 23.*

RATTLESNAKES

Ch-ch-ch-ch-ch-ch! A rattlesnake shakes its tail to say, "Move away!"

SERPIENTES DE CASCABEL

¡Ch-ch-ch-ch-ch-ch! Una serpiente de cascabel agita su cola para decir «¡Apártate!».

Hard **scales** form the rattle.

Duras **escamas** cubren el cascabel.

A **rattler** can shake its tail more than 50 times per second.

Una **cascabel** puede sacudir su cola más de 50 veces por segundo.

Rattlesnakes must shed
their skin to grow.

Las serpientes de cascabel
deben mudar su piel
para crecer.

Rattlesnakes add a ring to their rattles each time they shed their skin.

Las serpientes de cascabel agregan un anillo a sus cascabeles cada vez que mudan su piel.

Rattlesnakes may look tan, brown, or gray.

Las serpientes de cascabel pueden ser de tonos tostados, café o gris.

Rattlesnakes wait for **prey**.

Las serpientes de cascabel aguardan a su **presa**.

Pits on their snouts help them track their prey's body heat.

Las fosas del hocico les ayudan a rastrear el calor del cuerpo de sus presas.

Rattlers eat mice, birds, and other snakes.

Las cascabeles comen ratones, pájaros y a otras serpientes.

Sharp **fangs** squirt deadly **venom**.

Filosos **colmillos** chorrean **veneno** mortal.

These **reptiles** stay as warm or as cool as the air around them.

Estos **reptiles** se mantienen tan calientes o fríos como el aire alrededor de ellos.

Cold air slows rattlesnakes. They often warm their bodies in the morning sunshine.

El aire frío hace que las cascabeles se muevan más lento. Suelen calentar sus cuerpos con los rayos de Sol de la mañana.

Most rattlesnakes live in rocky deserts or forests.

La mayoría de las cascabeles viven en desiertos rocosos o en bosques.

Some people eat rattlesnakes. Would you?

Algunas personas comen serpientes de cascabel. ¿Tú lo harías?

Glossary/Glosario

fangs (FANGZ): Fangs are sharp, long teeth.
colmillos: Los colmillos son dientes largos y filosos.

prey (PRAY): Prey is an animal that is hunted by another animal for food.
presa: Una presa es un animal que es cazado por otro animal para alimentarse.

rattler (RAT-lur): A rattler is a nickname for a rattlesnake.
cascabel: Cascabel es un apodo para la serpiente de cascabel.

reptiles (REP-tilez): Reptiles are cold-blooded, scaly animals that breathe air.
reptiles: Los reptiles son animales escamosos de sangre fría que respiran aire.

scales (SKAYLZ): Scales are small pieces of hard skin that cover the bodies of reptiles and fish.
escamas: Las escamas son piezas de piel pequeñas y duras que cubren los cuerpos de reptiles y peces.

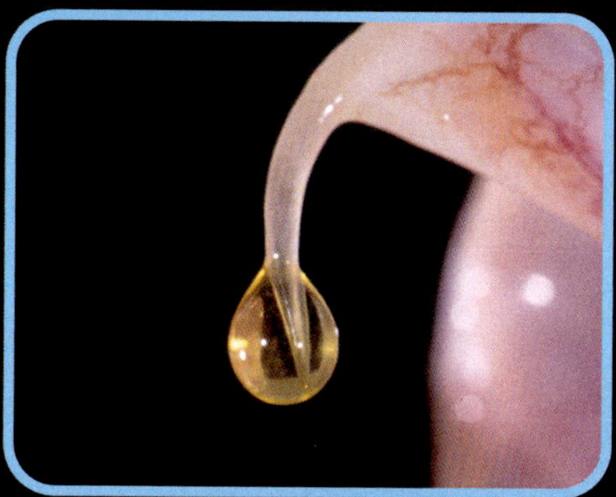

venom (VEN-uhm): Venom is a poison passed through a bite or sting.
veneno: El veneno es una ponzoña pasada a través de una mordida o picadura.

Index

Índice analítico

Websites
Sitios web (páginas en inglés):

https://kids.nationalgeographic.com/animals/reptiles/rattlesnake
https://easyscienceforkids.com/all-about-rattlesnakes

About the Author

Tracy Nelson Maurer has written more than 100 books for young readers. She lives in Minnesota where it's too cold for most dangerous snakes.

Acerca de la autora

Tracy Nelson Maurer ha escrito más de 100 libros para lectores jóvenes. Vive en Minnesota, donde hace mucho frío para la mayoría de las serpientes peligrosas.

Crabtree Publishing

crabtreebooks.com 800-387-7650

Hardcover 978-1-0398-6918-9
Paperback 978-1-0398-6897-7
Ebook (pdf) 978-1-0398-6939-4
Epub 978-1-0398-6960-8

Library and Archives Canada Cataloguing in Publication
Available at the Library and Archives Canada

Library of Congress Cataloging-in-Publication Data
Available at the Library of Congress

Published in Canada
Publicado en Canadá
Crabtree Publishing
616 Welland Avenue
St. Catharines, Ontario
L2M 5V6

Published in the United States
Publicado en los Estados Unidos
Crabtree Publishing
347 Fifth Avenue
Suite 1402-145
New York, NY 10016

Written by/Escrito por: Tracy Nelson Maurer
Designer/Diseñadora: Jennifer Dydyk
Editor/Editora: Kelli Hicks
Translated by/Traducción de: Sophia Barba-Heredia
Spanish-language copyediting and proofreading/Edición en español: Base Tres
Production manager/Coordinadora de producción: Candice Campbell

Photographs: mask for snakeskin graphic on cover and pages © shutterstock.com/Merydolla. yellow triangle with snake graphic © Top Vector Studio/Shutterstock; Cover photo: © Steve Byland/Shutterstock; page 3 © Audrey Snider-Bell/Shutterstock; page 5 © Creeping Things/Shutterstock; page 7 © Chaikom/Shutterstock; page 9 © Susan Schmitz/Shutterstock; page 11 © Ali Iyoob Photography/Shutterstock; page 13 © Breck P. Kent/Shutterstock; page 15 © Audrey Snider-Bell/Shutterstock; page 17 © Edgar Figueiredo/Shutterstock; page 19 © Brian Magnier/Shutterstock; page 21 © Brian Magnier/Shutterstock; page 23 (bottom photo) © Joe McDonald/Shutterstock

Printed in the U.S.A./CP062026